CHOPPERS

MOTORCYCLE MANIA

David and Patricia Armentrout

Rourke

Publishing LLC

Vero Beach, Florida 32964

www.rourkepublishing.com

PHOTO CREDITS: Cover, title page, pp. 7, 8, 9, 13, 15, 16, 17, 19, 20 ©thedigitalrose.com; pp. 5 ©Brian Ducharme

Title page: *Choppers are showy but powerful machines.*

Editor: Frank Sloan

Cover and page design by Nicola Stratford

Library of Congress Cataloging-in-Publication Data

Armentrout, David, 1962-
  Choppers : motorcycle mania / David and Patricia Armentrout.
    p. cm. -- (Motorcycle mania)
  Summary: "Climb aboard! Imagine cruising the countryside on a big, comfortable touring bike, rocketing down a strip on a drag bike, or gripping the ice in an extreme speedway race! If these excite you, then you are a motorcycle fan" "[summary]"--Provided by publisher.
  Includes index.
  ISBN 1-59515-452-3 (hardcover)
  1. Motorcycles. 2. Motorcycling. 3. Popular culture--United States. I. Armentrout, Patricia, 1960- II. Title. III. Series.

TL440.A7432 2006
629.227'5--dc22
                              2005012651

Printed in the USA

CG/CG

Rourke Publishing
1-800-394-7055
www.rourkepublishing.com
sales@rourkepublishing.com
Post Office Box 3328, Vero Beach, FL 32964

TABLE OF CONTENTS

FOR THE LOVE OF IT

People ride motorcycles for various reasons. Some love the flat-out speed of a tricked-out **drag bike**. Some crave the thrill of a race on a bare bones **speedway bike**. Others prefer a laid-back, cross-country cruise on a powerful touring bike. It seems there is a bike for just about everyone.

Motorcycle rallies are popular events in many parts of the country.

ONE-OF-A-KIND MACHINES

Motorcycles are built all over the world. Americans ride bikes made in Japan, Germany, Italy, the United States, and other nations. One kind of motorcycle, the chopper, is as American as apple pie. A chopper is a custom-made bike. Choppers are not made for racing or trail riding. True choppers are one-of-a-kind machines built for cruising the roads in style.

Bikers look forward to warm, sunny days so they can cruise the roads in style.

BOBBERS

In the late 1940s, after World War II, riders began to experiment with the design of their bikes. To make them lighter, fenders and other "unnecessary" parts were shortened, or bobbed. Some extra parts were removed altogether. The lighter bikes were easier to handle and fun to ride. Riders took pride in their creations. The new bikes were called bobbers.

A young admirer checks out a customized 1947 Harley Davidson.

Choppers have come a long way since the early days.

CHOPPERS

Riders continued to tinker with their bikes. By the 1960s, bobbers gave way to choppers. The name choppers came from the way riders removed, or chopped off, extra parts.

In the early days, before choppers became big business, most chopping, or **customizing**, was done by **amateur** mechanics.

Riders who don't have the skill or time to build a safe custom bike can buy one from qualified builders.

MODERN MARVELS

The choppers of today have **evolved** into complex machines. Customizing choppers has become big business. Many modern choppers are **engineered** from scratch. That is, they are no longer just chopped-up factory bikes. They are designed and built from the ground up by highly skilled craftsmen and women.

However, some bikes are still made the old-fashioned way, by riders who want to create their own masterpiece.

A sissy bar is the tall backrest behind the seat.

Parts for custom choppers are often built, or fabricated, from scratch.

ONE SIZE DOESN'T FIT ALL

Choppers are as unique as their riders. They come in many designs and sizes, but they usually have a few things in common. Most have an extended **fork**, no rear **suspension**, and high handlebars. Choppers often ride low to the ground.

High, long, handlebars are sometimes called ape hangers because the position of the rider holding the handlebars looks something like an ape hanging from a tree.

An American-style chopper with an extended fork and high handlebars

THE SKY IS THE LIMIT

Choppers tend to match the personality of their owner. Riders want their bike to fit their lifestyle. Some choppers have wild painted **graphics** and crazy looking designs. Others look like works of art.

Chopper designs are unique, but they all have one thing in common: They are almost never boring.

A big rear tire looks cool and gives a chopper good traction.

Many choppers have a small front tire and a large, fat tire on the back.

CUSTOM-BIKE BUILDERS

Custom bike builders are expert craftsmen. They shape and mold bikes to their clients' specifications. They have an "if you can dream it, we can create it" attitude when it comes to building bikes.

It all starts when a client makes a complete list of wants and needs for the bike. Clients choose the frame, engine, wheel and tire size, and all the extras that will make their bike unique. The builders go to work heating, pounding, and shaping metal into frames, fenders, and exhaust pipes. They sometimes polish up old parts, rebuild engines, and even reshape gas tanks.

Some clients simply have a theme in mind. They might ask for a patriotic bike or a bike that reflects their love for horses, fishing, or even snakes. They leave the design in the hands of the experts.

The owner of this chopper wanted a snake painted on his bike.

A POPULAR RIDE

In the 1960s, 1970s, and even the 1980s, choppers were not always well thought of. Many people thought only outlaw motorcycle gangs and troublemakers rode choppers. The truth was, most riders were simply motorcycle lovers who wanted a new style of bike.

Today, all kinds of people ride choppers. Doctors, plumbers, firefighters, and just about anyone who has an interest can ride a chopper.

Celebrity custom bike builders sign autographs at a Texas bike rally.

Bikers show off their rides in a motorcycle rally parade.

Choppers have become more common in recent years, but that does not mean they have lost their style. While the choppers of the past may have been unique, they were not always the safest motorcycles to ride. Skilled engineers have designed many of the new choppers to be more stable, safer to ride, and, of course, as stylish as ever.

Choppers give riders a way to display their personalities like no other machine on the road can.

There's no mistaking the low, throaty growl of a powerful chopper.

GLOSSARY

amateur (AM uh tur) — someone who works for pleasure rather than money

customizing (KUS tuh myz ing) — building something to a person's specifications

drag bike (DRAG BIKE) — motorcycles that are raced in an acceleration contest over a measured distance

engineered (EN juh NEERD) — designed by someone who is trained

evolved (ee VOLVD) — to have changed in appearance as something develops

fork (FORK) — the metal tubes holding the front wheel to the body of the motorcycle

graphics (GRAF iks) — images such as a drawing

speedway bike (SPEED way BIKE) — a racing motorcycle that has no gears or brakes

suspension (suh SPEN shun) — the system that cushions the motorcycle from dips and bumps on the course

INDEX

FURTHER READING

Hill, Lee Sullivan. *Motorcycles*. Lerner Publications, 2004

Morris, Mark. *Motorbikes: Mean Machine*. Raintree, 2004

Preszler, Eric. *Harley-Davidson Motorcycles*. Capstone Press, 2004

WEBSITES TO VISIT

www.ama-cycle.org

www.orangecountychoppers.com

www.dsc.discovery.com/fansites/amchopper/amchopper.html

ABOUT THE AUTHORS

David and Patricia Armentrout specialize in writing nonfiction books for young readers. They have had several books published for primary school reading. The Armentrouts live in Cincinnati, Ohio, with their two children.

32395